FOR

I THINK YOU'D ENJOY THIS BOOK BECAUSE

FROM

PRINCIPLES FOR THE NEXT CENTURY OF WORK

Sense & Respond Press publishes short, beautiful, actionable books on topics related to innovation, digital transformation, product management, and design. Our readers are smart, busy, practical innovators. Our authors are experts working in the fields they write about.

The goal of every book in our series is to solve a real-world problem for our readers. Whether that be understanding a complex and emerging topic, or something as concrete (and difficult) as hiring innovation leaders, our books help working professionals get better at their jobs, quickly.

Jeff Gothelf & Josh Seiden

Series co-editors **Jeff Gothelf** and **Josh Seiden** wrote *Lean UX* (O'Reilly) and *Sense & Respond* (Harvard Business Review Press) together. They were co-founding principals of Neo Innovation (sold to Pivotal Labs) in New York City and helped build it into one of the most recognized brands in modern product strategy, development, and design. In 2017 they were short-listed for the Thinkers50 award for their contributions to innovation leadership. Learn more about Jeff and Josh at www.jeffgothelf.com and www.joshseiden.com.

OTHER BOOKS FROM SENSE & RESPOND PRESS

Lean vs. Agile vs. Design Thinking
*What you really need to know to build
high-performing digital product teams*
Jeff Gothelf

Making Progress
The 7 responsibilities of the innovation leader
Ryan Jacoby

Hire Women
An Agile framework for hiring and retaining women in technology
Debbie Madden

Hiring for the Innovation Economy
Three steps to improve performance and diversity
Nicole Rufuku

Lateral Leadership
A practical guide for Agile product managers
Tim Herbig

The Invisible Leader
*Facilitation secrets for catalyzing change,
cultivating innovation, and commanding results*
Elena Astilleros

The Government Fix
How to innovate in government
Hana Schank & Sara Hudson

Outcomes Over Output
Why customer behavior is the key metric for business success
Josh Seiden

What CEOs Need to Know About Design
A business leader's guide to working with designers
Audrey Crane

OKRs at the Center
How to use goals to drive ongoing change and create the organization you want
Natalija Hellesoe & Sonja Mewes

Ethical Product Development
Practical techniques to apply across the product development life cycle
Pavani Reddy

To keep up with new releases or submit book ideas to the press, check out our website at www.senseandrespondpress.com.

HIRING PRODUCT MANAGERS

Copyright © 2020 by Kate Leto

All rights reserved. No part of this publication may be reproduced, stored in a retrieval system, or transmitted, in any form or by any means, electronic, mechanical, photocopying, recording, or otherwise, without the prior written permission of the publisher.

Issued in print and electronic formats.

ISBN 979-8-9853752-2-0 (KDP paperback)

Editor: Victoria Olsen
Designer: Mimi O Chun
Interior typesetting: Jennifer Blais

Published in the United States by Sense & Respond Press, www.senseandrespondpress.com

Printed and bound in the United States.
1 2 3 4   23 22 21 20

Kate Leto

HIRING PRODUCT MANAGERS
Using Product EQ to go
beyond culture and skills

CHAPTER 1: BUILDING A PRODUCT PRACTICE

In 2019, Pete was a newly appointed director of product at a global financial services firm, New Star. A stalwart of London's dynamic financial scene, New Star was in the midst of a multiyear product transformation that began in response to what the leadership team viewed as lack of transparency into what was being built, slow delivery of products to market, and an overall sense that without significant change in digital product strategy the market-leading bank would lose share to more nimble, digitally native competitors.

The director of product role was new to the bank, and Pete was the first leadership hire from outside to be brought in as part of the product transformation. His background seemed a perfect fit. He had spent the last 10 years at a competitor working his way up from associate product manager to head of product. He was known as someone who would never miss a deadline and went out of his way to make sure his teams delivered on time, on budget, and on scope.

Plus, his education was from a top university and he was personally recommended to the hiring executive, Sarah, by a mutual friend. Pete seemed the ideal candidate, and Sarah was very excited to welcome him to their organization.

When I first met up with Pete and Sarah at one of the bank's Central London offices, Pete was six months into his new job and the only context I was given before the session was that things were not going well. As the three of us introduced ourselves, the tension in the room was palpable. Sarah and Pete began by giving me some context of what had been happening inside the New Star product organization since Pete became director.

When he first joined, Pete, Sarah, and New Star's leadership team agreed that Pete would work toward three goals during his first six months as director:

» Deliver a prototype of a new mobile app for field-testing within three months. This was a big push for a group that hadn't delivered anything new to market—beyond website updates—in nearly a year.

» Upskill the team of 30 product people on best practices, ensuring they were using the most effective and progressive techniques.

» Hire up to 10 new product people, as the leadership team was confident more people were needed to deliver on their strategic vision.

Pete put everything he had into delivering on the new mobile app. To ensure the product people were properly tooled up and ready to move on the prototype project, he created a checklist that contained the step-by-step process and guidelines each team should follow to develop the new app.

In his mind, Pete was following through on one of his main goals and upskilling the product team. He backed up his how-to-list with an extensive archive of books, articles, videos, blogs, and podcasts that dissected the best practices on all things product development—from vision statements and roadmaps to user interviews, minimum viable products (MVPs), objectives and key results (OKRs), design sprints, and much more. It was a treasure trove of product techniques.

Pete was immensely proud of the detail he had provided for his new teams of product people and was convinced that if the teams would adopt these practices, they'd have no problem meeting their three-month goal.

But six weeks into his new role, he realized the teams weren't delivering. Instead of trying to understand why, he initially tried to pick up the slack and become more involved with their work, directing detailed steps and following up daily on progress. Pete considered this "hands-on" while the teams described it as "micromanagement." No matter how it was labeled, nothing positive was happening.

To try to get the project over the line, Pete took matters into his own hands and took the bulk of the work back from the teams. He had been working nonstop to try to get the work done on his own. Though he was able to tick a few things off of his checklist, at the end of the three-month timeline, there wasn't even a whiff of a prototype for testing. All that was accomplished was that Pete was approaching burnout and the team members were frustrated and confused and many were talking of leaving the company.

What went wrong? Sarah was sure she had found the "right" person for the job. Pete met all the requirements in the job description and his experience was perfectly aligned to that of New Star. Everything seemed like it should click. He had the right background and was on top of all the best practices that the industry leaders and influencers were touting to ensure his teams built a great product on time, on budget, and on scope. On the surface, New Star, Sarah, and Pete were doing everything right. Yet, there was minimal progress on any of Pete's goals, and the teams working with Pete were unhappy to the point of considering looking for new roles elsewhere.

Sarah said they were hoping that as a product and organizational design coach that I might have some quick tips to help them sort out the situation, get the teams moving again, and help ensure they had great hiring processes in place for the 10 new hires that they still were expecting to onboard.

Before responding, I took a slow, deep breath. The scenario is one that I had heard many times, and I knew there was no quick fix to offer them. "What you're experiencing is not uncommon and can be untangled." Smiles began to appear on Sarah and Pete's faces. "But," I continued, "it's going to take time and commitment to resolve. I'm afraid there are no quick tips to create the change you're looking for. In fact, from what you've shared with me so far, it sounds like your challenges relate to a fundamental misunderstanding of what product management actually is and how to hire for it."

Given that Pete and Sarah were responsible for leading the product management organization of a global financial services organization, my response wasn't what they expected. I found myself on the receiving end of two blank stares.

"Bear with me," I said. "Let me share my thinking on what product management is, and how it sounds like you're focused on only half the picture. First off, what product management is and what a product person does—from entry-level junior product manager to executive-level chief product officer—is continually evolving and often varies greatly from one organization to another. So, it's difficult to find a universal understanding of it." Pete nodded in agreement.

"Based on my two decades of product experience, the key to understanding product management is to think of it as a practice—somewhat like a doctor practicing medicine or a lawyer practicing law," I continued. "Product people select from a variety of tools that live in our virtual toolbox to solve a problem. Given that the technologies we're working with are often new, there's no sure way to solve that problem, so there's a lot of experimentation and trial and error. Sometimes it works, sometimes it doesn't—but teams of product people won't know for sure until they try. It takes a lot of practice, and it takes a special set of skills to be that person who can continually experiment in times of stress and pressure. It also requires a unique type of leadership and culture to empower teams to do just that."

The blank stares from Sarah and Pete began to ever so slightly soften, so I went deeper into the model and broke the practice of product management itself into two distinct yet complementary sets of skill: **technical skills** and **human skills**.

I described technical skills as a growing list of techniques and tools that product people use to solve the customer problem, and then deliver and support a product. "It's impossible to name all of the technical skills within a product management practice, as there are new iterations and approaches being blogged and tweeted about daily, but some of the most universally recognized include product roadmaps, vision statements, OKRs, key performance indicators (KPIs), discovery sprints, product prototypes, user interviews, plus how to work with various methodologies like Agile and Lean to ideally deliver a product."

Pete nodded knowingly as if he was the master of all of these skills, and many more.

I went up to the whiteboard on the conference room wall, and asked Pete to tell me a few of the technical skills he used as part the most recent prototype development project. He listed out things like a design sprint to help identify what would be included in the prototype, a vision statement for the idealized product, a roadmap for development, and some interview techniques he was planning to use when testing the prototype with customers. "Anything else that you incorporated into your work," I asked?

"Nope, for this project that was about it." He replied confidently.

"This is actually a perfect example of where things went off the rails for you and the prototype project," I said. "In the technical skills dimension of a product practice, these—and many other tools—represent *what* a product person does as part of the ideation, creation, delivery, and iteration of digital products. They are essential activities for you, your team, and any product person. We as product people spend *a lot* of time mastering them."

"But this is really only half of what product management is all about," I continued. "What a product practitioner at any level should actually be working toward is the ability to balance technical skills with human skills. The ability to influence, lead, coach, mentor, learn, deal with conflict, communicate, be curious, innovate, adapt, energize, build and maintain relationships, and continually grow as an individual. These skills describe *how* a product person works and must go hand-in-hand with the technical skills."

I went on to describe how human skills are different from technical skills, yet the two are inexorably tied together for the product person. "For example, say a product manager has been tasked with building a new vision for a product. Successfully creating a vision statement isn't just knowing the latest frameworks or doing what the highest paid person in the room (the HIPPO) says; it's having the ability to influence challenging stakeholders to get alignment on your thinking. It's displaying collaboration and leadership skills to bring your team along with the product vision. It's conflict resolution skills, because we all know there is—and always will be—conflict and tension in product work, especially around something as debatable as a vision statement."

Sarah and Pete shared a brief smile, and Sarah went on to say that they definitely were up against HIPPOs in the now-defunct prototype project and that both she and Pete could have used some help influencing the various executive stakeholders who were pushing their own agendas and options.

"Human skills are as essential to the product practice as are technical skills," I continued, "and achieving this balance is perhaps more important now than ever. While the technologies we build upon are quickly becoming ubiquitous, it's the human approach to creativity, innovation, decision-making, and leadership that makes the difference in whether an individual, team, product, and even organization is successful or not."

I went up to the whiteboard on the conference room wall, and added a few examples of technical and human skills to provide the full picture of what a product practice is:

PRODUCT PRACTICE EXAMPLES

TECHNICAL SKILLS: *WHAT* WORK IS DONE	HUMAN SKILLS: *HOW* THE WORK IS DONE
» Product roadmaps » Vision statements » OKRs and KPIs » Design sprints » Product prototypes » Testing with customers » A/B and multivariate testing » MVPs	» Influence » Leadership » Active learning » Resilience » Adaptability » Creativity » Dealing with conflict » Emotional intelligence (self-awareness, self-management, social awareness, relationship management)

"Just know that you're not alone, and the situation you're finding yourselves in isn't surprising," I added. "In fact, the vast majority of product people and organizations that I've worked with base hiring decisions and advancement on a candidate's ability to master technical skills, much to the detriment of human skills and building a balanced product practice. From what we've talked about so far, it sounds like you are falling into this human skills gap."

I then drew a chart on the board to show an estimation of their product practice today and where we would ideally move to. "The outcome of any work that we might do together would not be to take focus off of technical skills, but to create a more balanced product practice by integrating human skills into ways of working and hiring. I drew a pair of simple pie charts to illustrate the change that we would work toward.

WORKING TOWARD A BALANCED PRODUCT PRACTICE

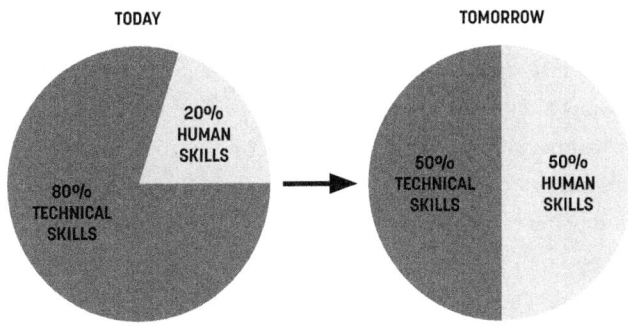

It is possible to grow and evolve our human skills, and that's what Pete, Sarah, and the teams at New Star needed to focus on. They needed to commit to learning more about human skills and how their own behaviors can have just as big of an impact—if not bigger—on a product, a team, and an organization than any of the latest product techniques.

Shortly after our initial meeting, I was asked to work with Pete, Sarah, and the product teams to help them begin to realize these changes. Based on previous experience, I knew that our work would be focused on a few key areas:

» Understanding what a balanced product practice is and why hiring for human and technical skills sets is crucial
» Applying this learning to transforming how new roles are created and interviews are conducted
» Building on human skills to rethink hiring for organizational fit and diversity
» Creating opportunity for ongoing growth in the hiring process by integrating continuous learning and reflection.

As I continue to share the New Star story, I'll describe practices from within these areas of focus that you, your team, and the wider organization can experiment with immediately to evolve your own hiring process by building and integrating human skills.

I'll also share **activities** to help you reflect and grow personally along the way. As you work with human skills, you may find that many of the lessons applied to your hiring processes can also promote personal growth. The activities are featured at the end of the chapters and are based on work done with Pete, Sarah, and members of the product team in individual and group coaching sessions.

The product organization was the start of a piece of work that would transform how product management was thought of and hired for at New Star. Surprisingly to Sarah and Pete, the first step in our journey existed in an unlikely place—the job description. For many, creating a job description is just a knee-jerk reaction to discussions on hiring someone new. However, how a job description is created, the language used in the description, and the requirements outlined provide great insights into how a company really thinks of the position. If designed well, the job description can lay the groundwork for building or enhancing a great team.

As Sarah was the hiring manager for the director of product role, I went to her for more context as to how the role was envisioned, what went into the decision to hire at that level, and the type of human and technical skills they were looking for in the role for which Pete eventually applied.

ACTIVITY: SELF-REFLECTION ON YOUR OWN PRODUCT PRACTICE

To kick off our work together, I asked Pete to do some homework. I wanted to help him build an understanding of where his product practice was when we started to work together and how he can start to create change. I asked Pete to find 15 minutes to do a self-reflection on his product practice by answering a few questions.

» *On a scale of 1 to 10, how would you characterize your focus on technical skills and human skills in your own product practice?*
Start by identifying a number from 1 to 10 for technical skills and also for human skills. For example, a recent client scored themselves an 8 out of 10 for technical and 6 out of 10 on human skills. This initial score should be a general representation of where you think you are. It's just a place to start, so don't overthink it. I shared a list of technical and human skills with Pete for inspiration that can be found in the back of this book.

» *What is holding you back from bringing the two dimensions more into balance?*
For example, what is your own awareness of or comfort level with human skills? Does your team or product organization encourage development in technical skills but not human skills?

» *What small changes could you make, or experiments could you put in place, to improve your score by one point?*
How can you get that 6 on human skills up to 7? For example, if you'd like to focus on increasing

self-awareness, can you challenge yourself to ask five colleagues for feedback in the next two weeks? If you'd like to work on your ability to influence, can you push yourself to set up meetings with two tricky stakeholders to understand their own challenges or concerns on a recent project?

Feel free to answer these questions yourself and share your responses with someone. It could be a peer, mentor, manager, or even your wider team. The more comfortable you are sharing your work on human skills, the more others will want to be involved and the better chance you have of actually following through on your personal challenge.

Pete shared his initial scores with me. He scored his comfort level with technical skills at an 8 out of 10 and his human skills at 4 out of 10. He knew he had work to do on human skills and felt that focusing on building self-awareness was a first step to understanding his own actions and behaviors and how they were impacting those around him—especially the product teams he was leading.

As an initial experiment, Pete challenged himself to ask for feedback from five members of the product teams over the next week and would share that feedback with me in our next session.

UNDERSTANDING EQ AND HUMAN SKILLS

Before continuing with the New Star story, I'd like to take a moment to talk about three key concepts that appear throughout this book: human skills, emotional intelligence, and Product EQ.

The human skills discussed in this book are often talked about in relation to the broader concept of emotional intelligence. Similar to product management, the definitions, frameworks, and competencies associated with emotional intelligence can get a bit complicated. A good basic definition comes from the Institute for Health and Human Potential, which describes emotional intelligence as:

> *The ability to recognize, understand and manage your own emotions and the ability to recognize, understand and influence the emotions of others.*

The term *emotional intelligence* was first coined in 1990 by two Yale University professors, John Mayer and Peter Salovey, and was expanded and popularized by psychologist and author Daniel Goleman's 1995 book, *Emotional Intelligence: Why It Can Matter More Than IQ*.

In the years since Goleman's publication, the concept of emotional intelligence has evolved into numerous schools of thought with various models and frameworks that are often combined and referred to as EQ, which is the abbreviation of emotional quotient (how a person's emotional intelligence is measured.)

An in-depth look at emotional intelligence and EQ could dominate the pages of this book, but for the purposes of our ongoing look at New Star, let's take a brief look at the basic dimensions as defined by Goleman:

- **Self-awareness:** The ability to know what we're feeling and why we're feeling it; self-awareness is the basis of good intuition and decision-making.
- **Self-management:** The ability to handle distressing emotions so that they don't cripple us while also being able to connect with positive emotions to get involved with and enthused about what we're doing.
- **Social awareness:** The ability to handle relationships and awareness of others' feelings, needs, and concerns.

» **Relationship management:** The convergence of the other three competencies; relationship management is more than ensuring relationships are maintained, but that they are also positive and beneficial for both parties.

These competencies can be tied to the human skills within a product practice and how work—within or outside of product management—is done. **However, the concept of Product EQ that I have developed doesn't need to be restricted to in-depth understanding of emotional intelligence frameworks and research.** Just as the practice of product management continues to evolve, so does our understanding of the human competencies and skills that shape how we work.

If you're interested in reading more specifically on emotional intelligence, check out the **Reading List** in the back of the book.

CHAPTER 2: DECIPHERING THE JOB DESCRIPTION

Sarah didn't come from a product background, but she had been involved in the bank's project management function for many years. In one of our first one-on-one sessions, she described her experience with product by saying that she didn't have a lot of hands-on experience, but that she had the "product mindset" needed to understand the work and lead it going forward. She then went on to walk me through a presentation that she had given to the board about a five-year strategy for digital products.

Sirens started going off in my head. Anyone who says they have a "product mindset" to lead a product organization but then goes on to talk through a five-year product plan is lacking the awareness needed to realize that ideas today will not meet the ever-changing needs of a product company in five years. A plan of this nature is more of an exercise in putting fantasies to paper than a worthwhile use of time. It's another great example of the importance of human skills in a product practice, but that was a lesson for another day.

Getting back to the topic at hand, I responded, "OK, if that's where you want to go, how are you going to get there?" Sarah immediately started talking about the importance of the new director position—Pete's role—and said his was a key hire to figuring that out.

As I pushed for more information on why so much focus was placed on the director role, for what the role was designed to be accountable, and what type of thinking Sarah and other members of the team brought to the role, Sarah looked at me blankly and handed over the job description for Pete's role.

My heart sank—getting a response to basic questions about the design and vision of a role in the form of a job description is not a good sign. It tells you right away that the organization hasn't thought about what the role will mean to a transforming product organization or the type of person this role is designed for. Job descriptions—especially for roles in areas as varied and dynamic as product management—often fail to encapsulate the true purpose, accountabilities, and technical and human skills essential to the role.

I looked at the job description, and it was comparable to others I'd seen for similar roles. It was based on the generic template of requirements for the job that are "must-haves" and those are that are "nice-to-haves."

The "must-haves" often detail technical skills the company believes are a requirement for any candidate. For example:
- » Experience in a specific type of market (B2B or B2C), industry (ecommerce, healthcare, financial services), or technology (AI, big data, biometrics)
- » Experience in certain types of technical activities: creating strategy, vision, OKRs, design sprints, MVPs, etc.
- » A specific level of education is occasionally listed, with an MBA still being favored by some product organizations.

Meanwhile, the "nice-to-haves" usually present the human skills that an organization believes are required for the job. When it comes to product management job descriptions—from junior manager to vice president—this section often starts with a common, undescriptive statement that it would be nice if the candidate "deals well with ambiguity" and then goes on to call for qualities like "good communicator" and "collaborates well."

As with many product job descriptions—New Star's list of must-haves at the top was twice as long as the nice-to-haves at the bottom, and the language used throughout both sections painted the picture of an ideal candidate who knew it all, had done it all, and possessed fantastical human qualities. The description included:

Must-haves (technical skills)
- » Proven experience managing business-to-consumer product teams
- » Proven track record of managing all aspects of a successful product through its life cycle

- » Proven experience developing new product strategy and communicating it to executives
- » Proven technical background with hands-on experience in software development and web technologies

Nice-to-haves (human skills)

- » Comfortable with the ambiguity that comes from doing new things
- » Natural-born leader; people want to work with you
- » Anti-fragile; you bounce back after failure
- » Self-starter: a driven, motivated individual who identifies opportunities, is prepared to get their hands dirty, digs into details, and pushes things over the line.

It's no wonder that this job description appealed to Pete—who in just a few months' time displayed his know-it-all approach to leadership and motivation to get things over the line—and in so doing alienated the product teams and brought work to a standstill.

I asked Sarah how the description was created, and she responded with the shrug of her shoulders. "Someone in HR" was her best guess.

"The teams of product people and stakeholders who would be working with Pete didn't have any input?" I asked. "No," responded Sarah. "That would have taken too long to pull together, and the HR people know what to look for."

I later spoke to a few of the HR recruiters involved in Pete's hiring and found that the job description was based on a collection of other descriptions they had found online. It was an amalgamation of product roles from other banks and industry leaders like Google and Amazon. This type of job description creation by cutting and pasting isn't surprising to hear about, nor unique to New Star. In fact, a recent survey of HR hiring managers found that though 80 percent say that job descriptions are important, about 50 percent admit to cutting and pasting together job descriptions.

The lack of thought, effort, and consideration of a role that was so crucial to delivering a long-term strategy for New Star once again confirmed that Sarah and New Star had a massive gap in understanding what product management was, let alone how to hire for it. Before the team jumped back to a world of must-haves and nice-to-haves for their next hire, we needed to shift thinking from a tactical job description to a meaningful, relevant role that focused on technical and human skills.

CHAPTER 3: USING THE ROLE CANVAS

We started out by holding a role creation workshop for the new senior product manager position that New Star was looking to add to one of its teams. The session was attended by all the members of this cross-functional team—including the current product manager, engineers, designer, and data analyst. Pete, Sarah, and the group's recruiter also joined—about 10 people total. To help facilitate the session, I introduced a tool called the Role Canvas.

ROLE CANVAS
PURPOSE: Why does this role exist?
ACCOUNTABILITIES: What are the goals or outcomes the role will be working toward?
HUMAN SKILLS: Leadership, conflict resolution, influence, adaptability, etc. / **TECHNICAL SKILLS:** Roadmaps, design sprints, product vision statement, JTBD, OKRs, etc.

© Kate Leto

Download the Role Canvas at www.KateLeto.com

The Role Canvas is based on four fundamental questions. By answering each question, a new piece of the role becomes clearer for the hiring manager, team members, stakeholders, and recruiters involved.

1. What is the purpose of the role? This isn't simply restating the job title, but goes to the core of why the role really exists and what the role will be working toward every day. For example, the purpose of the senior product role may be to lead the team to find new ways to engage and increase profitability with a new millennial market.

2. What is the role accountable for? What outcomes and goals will the role be working toward to deliver on its purpose? Driving usage with a target audience? Launching new feature or service?

Hitting a new level of customer service rating? It's often challenging to know what the role will be working toward in the longer term—or even in nine to 12 months. Start with the known goals or outcomes. Another way to approach this question is by flipping it, and first answering what this role is *not* accountable for. Depending on the way your team describes your work, you can also think in terms of what the role is/is not responsible for.

3. What human skills will the role need to display to achieve outcomes? For example, if this role is going to be part of a team that has experienced a lot of tension or conflict recently, the person who takes the role needs to have strong conflict resolution skills. Or if the team is working with challenging stakeholders, the role needs to be filled by someone with strong influence or even leadership skills. List the human skills needed for the role to be successful and prioritize them (see the list of human skills in the back of the book for inspiration).

4. What technical skills will the role need to execute in order to meet the outcomes and achieve success? Will this role need to create a product vision or strategy? Use A/B testing or lead a design sprint? Build a new roadmap? Use Jobs-to-be-Done? List all the technical skills the role will need to draw upon here and prioritize (sample technical skills also listed in the back of the book).

Facilitating the group through these seemingly simplistic questions created a lot of debate and dialogue. Our goal for the session was to come up with a first draft of a role canvas that featured the group's thinking on each question, and with the aid of the product manager's favorite props—sharpies, stickies, and stickers for dot-voting—we were able to do that.

The team left the session with a completed canvas that represented their communal, current thinking on the role—but knew there would be many iterations to come. As the team began to meet with candidates, their perspective about the role's purpose, accountabilities, and human and technical skills became clearer. The canvas is designed to be dynamic—the team updated it as they learned more about what they were and were not looking for.

Over the next few weeks, the team firmed up the role canvas and translated it into a job description that they felt represented what they were looking for and best described the opportunity for someone to join the organization.

ACTIVITY: CREATE A CANVAS FOR YOUR ROLE

What would a canvas for your own role look like? Before taking part in building a canvas for a new role, take this opportunity to reflect on your own.

- » Use the canvas as a tool for self-reflection and answer: What's the purpose of your role? What are you being held accountable for? What human skills do you feel your role calls for that you could improve on? What technical skills?
- » Share your thinking with members of your team and your manager to start a conversation about how you see the work that you're doing now, and where you could grow. Are your team members or managers

surprised by any aspects of your role? Can they help you identify gaps or similarities with their own roles?
» Revisit your role canvas quarterly or biannually and reflect on how you and your role have changed. Remember, roles are dynamic and evolve; use your role canvas as an aid to reflect and foster conversations about what the change may mean to you, your team, and the wider organization.

CHAPTER 4: INTERVIEWING FOR HUMAN SKILLS

Once the description for the new senior product manager role was posted, the hiring team, Sarah, and Pete started to schedule interviews. I asked about how the team was preparing for the interviews, and what plans they had. I was told that it was all under control. They would prepare as they always had: by writing some questions focused on the candidate's background and most recent role, and then doing a quick search at job and recruiting sites like Glassdoor.com for brainteaser questions from the likes of Google or Amazon that are sure to stump the candidate/victim.

Pete said that during his interviews with New Star, he was asked "What has more advertising potential in Boston: a flower shop or funeral home?" and to "Design an evacuation plan for the building."

What did you think of the questions? I asked Pete. "I thought they were completely random, and really didn't have anything to do with my past accomplishments or the job I was there to talk about," he responded. "To be honest, I wasn't sure what they were trying to achieve with the questions, and it made me stop and think if this was really the place for me."

I was on Pete's side, and was shocked by the questions the team thought were appropriate for someone interviewing for a senior product manager role. Designed to see how a job candidate responds to pressure and how they handle ambiguity—as stated in the nice-to-haves on the job description—these questions may make the interviewer feel clever, but they don't do much to help you understand if the person you're talking to would fit your role or organization.

I went on to tell the team about behavior-based questions, which are seemingly simple interview questions that go beyond the brainteaser or basic inquiry into technical skills and accomplishments that are often the focus of product interviews.

We talked about basic questions they often ask when interviewing for technical skills. For example: What products have you delivered? What tools (e.g. technical skills) did you use? Did you deliver on time and on budget? What were the metrics and results?

Diving deeper, behavior-based interview questions focus on understanding the behaviors that led to those accomplishments, the intentions behind the behaviors, and consequences and impact of the behaviors on others. According to a study by Ann Marie Ryan and Nancy Tippins from Michigan State University, behavior-based questions have the highest validity of all interviewing tools when used in a structured format.

They also provide the interviewer an opportunity to gauge body language and tone as the candidate responds to questions about the times when outcomes and results didn't meet intentions, which can be extremely telling.

Behavior-based questions can be tied directly back to the human skills listed in your role canvas. For example, if your canvas includes a requirement for someone with high levels of conflict resolution skills as a priority, ask questions that help you understand how a person has previously acted in times of tension and conflict at work. In her book *The EQ Interview: Finding Employees with High Emotional Intelligence,* Adele B. Lynn lists examples including:

- » Tell me about a time when you suggested something that someone disagreed with? What did you say?
- » Have you ever encountered someone at work who was unreasonable? What did you do?
- » Tell me about a time when someone felt that you were unfair. What did you do?

Looking for someone with a high level of resilience? Lynn suggests trying these prompts:

- » Tell me about a time when you decided to give up on a goal.
- » Tell me about a time when you were distracted or preoccupied at work. What did you do?
- » Tell me about the last time you were criticized at work: How did that go?

These are basic questions that aren't designed to trick, panic, or confuse the person who has taken time out of their own busy schedule to come to speak with you about joining your team. Instead, these questions give the interviewer an opportunity to hear how a person has behaved in the past and identify patterns that will build a picture of how—without any changes in behavior or intentions—this person would likely behave in the future.

With behavior-based questions, the burden is on the interviewer to really listen to what the candidate is saying and to be able to identify the intentions behind a behavior by probing and digging further into the narrative until they can make a seemingly intangible skill like conflict resolution tangible and answer questions like:

- » How does this person respond to conflict?
- » Do they walk away or engage?
- » If they engage, how?
- » Do they want everyone involved to walk away from the encounter feeling good or are they looking for a clear winner or loser?

A key point with behavior-based questions is to remember that though they can help us understand how someone has behaved in the past, we need to also recognize that with a commitment to change, people can evolve their behaviors going forward. Responses to behavior-based questions will give you a much better perspective of the candidate's human skill capabilities than asking about the advertising potential of flower shops in Boston. And perhaps you'll even be able to identify the candidate's desire to learn from the experience and change.

Before the first round of interviews, Pete and the product team that was leading the search for the senior product manager met to talk through the human skills they prioritized on their role canvas. They spent time brainstorming behavior-based questions that would help them understand the candidate's intentions and thinking behind their behaviors.

They also made a commitment to come back together after each round of interviews to share their experience with the questions and the responses they received in what we called a calibration session. The session was specifically designed to give the team an opportunity to iterate on the interview structure and questions they had devised to help them better evaluate human skills essential for the role. It was also a space to share their thoughts on how the interviews in general were going and what, if any, changes they should make moving forward.

With the initial questions identified, interviews scheduled, and first calibration session on the schedule, it was time to start interviewing.

ACTIVITY: PERSONAL INTERVIEW

Take time to reflect on how you would respond to the behavior-based questions you're about to ask a candidate. This can be an individual reflection, an exercise with a peer, or even with your team as part of preparation for interviewing.

» What lessons can you take from your own responses? For example, thinking about your response to a question about conflict: "Tell me about a time when you suggested something that someone disagreed with." Was your recent reaction to that scenario to storm out of the room? Ask yourself how you would have preferred to respond to the situation.

» What small step can you take to help ensure the reaction will be one you want? Is it challenging yourself to count to five silently before reacting? Taking three deep breaths?

» Find an experiment that works for you. How does understanding your own responses form a better understanding on what type of response you would look for from a candidate?

MEETING CHANGE

A week after interviewing for the senior product manager role began, I sat in on the hiring team's first calibration session. Jenny, one of the key members of the interviewing team and a five-year veteran of New Star, brought up something that she'd noticed over the past week. "The people I've met so far are very smart, and for the most part fit our technical and human skill profiles. They seem like really nice people, but I'm having a hard time seeing how they will mesh with our teams now."

"Has anyone noticed that the type of person that we're talking to now is very different from what we were looking for even a month ago?" she continued. "We never would've looked for someone with conflict resolution skills or organizational awareness previously." A few others in the room nodded in agreement.

"Can you tell me a bit more about the difference that you're noticing?" I asked. "Are you concerned that the human skills you identified in the role canvas don't really represent what you need?"

"I don't think that's it," Jenny responded. "I think what we're looking for now is good. It's just taking us in a different direction than the type of people we usually meet with. Which I know is what we want to do. I'm just wondering when we hire someone to fit our role canvas, will they fit into our ways of thinking and working now?"

Tom, another member of the hiring team, spoke up: "Maybe we need to take another look at our onboarding process to make sure that this new person feels welcome."

"I think it's more than having a slick onboarding process," continued Jenny. "Knowing that this is just the first new role we're looking for, and there are more to come, I think what we're building toward are some bigger, unexpected changes; ones that will have a bigger impact on our ways of thinking and doing."

"Maybe I'm thinking too far down the road, but has anyone really thought about the impact that these new additions will bring to our culture?"

The room was quiet. You could've heard a pin drop. I stood in the back and just smiled.

Jenny's realization was the first sign to the team that their work with human skills had the potential for a much greater impact on the organization than just a new hiring process. The seemingly small steps that they had taken so far to introduce human skills, and integrate them into their own personal practices and the new roles they were creating, were indeed shifting the group's thinking about what they were looking for in a new team member. But, shifting thinking and actually meeting people that embody that change are two very different things.

Through the past week of interviewing, the team for the first time was actually meeting people who possessed these human skills they now were identifying as desirable; and they were different from the norm at New Star.

As I looked around the room, I saw looks of fear, confusion, and even excitement. The work we had done so far was not going to just change their hiring processes; it was going to change the makeup of their teams, how they worked, the products they produced, the culture they were all a part of.

I shared with the team a simple, yet classic model to describe this sort of domino effect that they were starting to see. It's called the 7S Framework and was developed by Tom Peters and

Robert Waterman, consultants with McKinsey & Company in the 1980s. The framework identifies seven elements of an organization and asserts that changing, pulling, or pushing on one of these elements will create change in all others.

7S FRAMEWORK
- » Strategy
- » Structure
- » Systems
- » Shared values
- » Skills
- » Style
- » Staff

Keeping in mind that the terms used in the framework may be out of date—especially "staff," which we referred to as "people" in our conversation—the underlying sentiment aligned with what we were starting to see at New Star. By changing how New Star thought of the skills it now described as essential to its teams—human as well as technical—it was impacting the style of work, the systems the teams were operating within, their shared values, and more. With each role canvas, with each interview, the teams at New Star were redefining their culture.

We had initiated a chain reaction.

HUMAN SKILLS IMPACT

INDIVIDUAL ⇨ TEAMS ⇨ ORGANIZATION

Pete spoke up: "I remember in our first meeting with Sarah, you said there was no quick fix to our situation, Kate, and that we had to fundamentally change a lot of our thinking, and that it was going to take time."

"Yes, I remember that." I responded.

He cracked a smile. "Is this what you had in mind from the beginning? That changes to how we hire would actually create changes to our culture?"

I just smiled and said, "You got it."

Our first calibration session was coming to an end. It was a big, revealing conversation—one I don't believe the group expected to have when they stepped in the room—but one that gave me a very good feeling about the team becoming more aware of how their work on human skills could impact not just how they hire but who they were as a team and organization.

Before leaving, I asked everyone to go around the room and share one word that captured how they were feeling about the conversation today. I wrote the words shared on the whiteboard.

» Excited
» Scared
» Positive
» Unsure
» Hopeful
» Intrigued

They asked me for a word as well. "Elated!" I said with a big grin. I knew we were headed in a great direction.

"You can tell a lot from the question you asked at the end of the session, asking for one-word that captured how they were feeling about the conversation," Pete said as we walked out of the meeting room.

"It's a great practice that can be used at all meetings," I responded. "It's called a check-out round and is used best when you match it with a check-in round at the beginning of the meeting. In both instances, a simple question is asked to bring the group's attention to the conversation or provide a moment of reflection. Each person is invited to respond without cross-talk or dialogue."

I mentioned that I've been experimenting with using the practices in sessions over the past week and getting a great response. Pete and I agreed it would be interesting to bring them into our sessions with Sarah and even our one-on-one coaching sessions so that he could get more comfortable with the practices as well. To give him a better understanding of the practice, I promised to send some information over later in the day.

The next day, Jenny sent me an instant message, saying that she had been thinking about the recent session and had another big question: "In the past, we've thought that hiring for cultural fit was important. If our culture is changing though, where does that leave us in trying to asses if someone is going to fit in or not at New Star, or even know what cultural fit means to us anymore?"

"Those are great questions!" I responded.

"I'm glad you think so," Jenny said, "because a few of us from the hiring team were talking all about it this afternoon."

We agreed to bring the topic to our next calibration session so that everyone could take part in the conversation.

ACTIVITY: CHECK-INS AND CHECK-OUTS

Introducing check-ins and check-outs to Pete was a first step in transitioning from his own learning about human skills to building a sense of awareness and reflection with others through a practical exercise. Here's a simple facilitation guide to try it at your next meeting:

Check-in: Our days are often packed with meetings and things to do. A check-in round can quickly bring attention to the meeting or conversation at hand.

- » At the beginning of the meeting, ask each person a simple question to gauge participants' energy and presence. Typical questions: What has your attention? What will you bring to today's meeting? What is one thing that you'd like to accomplish in today's meeting? Or, on a more lighthearted note to build energy in the session, you could try something like: What's your favorite time of day? If you could be anywhere in the world for five minutes and then automatically return, where would you go?
- » Give the group a moment to consider the question and then invite someone to kick off the responses.
- » When someone is responding to the question, they have the floor. There is no cross-talk or conversation about the response. This allows for a more inclusive mindset to the meeting as everyone is invited to speak in turn.
- » The check-in should take no more than five minutes of the meeting. For larger groups, ask for a one-word or one-sentence response.
- » The check-in is not designed to be a platform for a monologue from team members, but a short exercise to help each person feel they are seen and heard and focused on the conversation at hand.

Check-outs: This is a similar exercise held at the end of the meeting to emphasize reflection and closure of the conversation.

> » The facilitator again poses a question to the team, for example: What did you notice in today's session? What did you learn? How are you feeling about the outcomes from today's meeting? What's your number-one takeaway?
> » Each person again responds to the question, without cross-talk or dialogue.

Check-out responses can again be limited to one word or one sentence and should take up five minutes (or less) of your meeting time.

CHAPTER 5: GOING BEYOND CULTURE FIT

When we gathered a few days later, I asked the team to tell me what they saw as the advantages of hiring someone who you think will fit into the New Star culture. I wrote the responses on the whiteboard:

- » People that fit in just gel with everyone easier. It's like having another friend at work.
- » Decisions are made faster because we usually agree on what we should be doing.
- » There's less tension and debate.
- » If they fit in, they seem to stay longer at the company. If they don't fit in, they tend to leave after a year or two.
- » Work gets done faster because we're all thinking alike.

I nodded my head and talked about the research behind the power of hiring for cultural fit and the advantages that it brings to organizations, team members, and even the products produced. According to a paper in the *International Journal of Business and Social Science* entitled "Ten Ways of Managing Person-Organization Fit (P-O Fit) Effectively," when a company's values, norms, culture, and goals align with a person's personality, values, attitudes, and goals, fit is actually achieved. When organizations find fit with a new team member, they can experience a drop in turnover and an increase in performance. Team members may also be better equipped to deal with change, a significant benefit for an organization like New Star going through transformation.

I asked about the type of candidate that they would've looked for before we started working together and what that fit would've looked like. Jenny smiled at Pete, her new boss, and she said, "Well, someone like Pete. He went to a great school, worked for a competitor, did well there, and we all knew a lot of the same people in the industry—his background is what we have always looked for and believed would be a perfect fit for our culture."

However, we were all aware that six months into Pete's time at New Star, members of his team were threatening to leave the company and development was at a standstill. His style of work and leadership clearly didn't bring the expected benefits.

"There's no doubt that there are advantages to hire for a fit," I said to the group. "But, can anyone think why it might *not* be a good idea to hire for fit?

Jenny piped up: "Well, pretty much all the same reasons that we've already listed above. We think alike, we make decisions really easily without debating them, we have the same experiences. We all get along, which is great, but we don't really stray from our norms or challenge each other."

"Exactly." I replied. While hiring someone for fit has its benefits, it's something we need to watch very carefully. In our exuberance to find what we believe is someone who will fit in, we often overlook different cultures, lifestyles, and ways of thinking that are actually really important to creating a more diverse team, and better products.

I shared a *Harvard Business Review* study, "How Diversity Can Drive Innovation," that found companies with more diversity in leadership out-innovate and outperform those with less diversity in lead roles.

The study's authors, Sylvia Ann Hewlett, Melinda Marshall, and Laura Sherbin, defined diversity in two ways:

» Inherent diversity—traits that you're born with like gender, ethnicity, and sexual orientation
» Acquired diversity—traits you gain from experiences like working in another country, which can give you an appreciation of another culture, or creating products for female consumers, which can give men some insights into women.

Their research showed that organizations with greater levels of both inherent and acquired diversity were 45 percent more likely to report an increase in market share over the previous year and were 70 percent more likely to capture a new market.

Organizations with inherent and acquired diversity encourage divergent thinking, they desire debates that push and pull apart how and what decisions get made, and they create opportunities for curiosity and innovative thinking.

"So cultural fit is important, but if we put too much focus on it, we're going to just build a team of clones and most likely just keep doing what we're doing now," remarked Jenny.

"Yes, *and* I think the challenge in front of us is actually bigger than just keeping an eye on how much importance you are putting on fit in hiring," I responded. The real challenge is changing thinking about what cultural fit really means. It's not just what makes up the New Star culture now—it's understanding what's missing from your culture that can help you build for the future.

I mentioned Jeff Vijungco, former VP of employee experience at Adobe, who said that when he took a look at a team, he often questioned whether it felt or looked like a stack or like a puzzle. Why? Because puzzle pieces are different, but complement each other to form a whole. A stack is just more of the same. Instead of cultural fit, Vijungco and Adobe looked for "culture complements."

IDEO, the global design firm, looks for "culture contributions" that someone can bring to their organization that they don't currently have, not just someone else that can fit into their existing ways of thinking. Spotify has a similar practice of going beyond fit to a "culture add-on."

Pete chimed in: "So no matter what words we used to describe it, we should actually find a way to evaluate if a candidate offers a new dimension to the culture that might be missing, like a missing piece of a puzzle; not just replicating what we already have."

"You got it," I responded. "There are few things we can do to put things into motion. The first it to lay a foundation by taking a look at our culture now and seeing if we're aligned in thinking about what the New Star culture is, what our values are, and what they actually mean to us. No matter if we're hiring for culture fit, adds, or contributions; we want to find people who have that underlying alignment with New Star's values."

"Core values make up the fabric of any company's culture and are fundamental principles for how work is done. They also signal the type of human skills we feel are important in our team members and what is missing. But, at the same time, we have to remember that commonality in values does not mean hiring clones that look, think, or operate as you do now," I added.

To collaboratively and collectively put some shape to the product organization's values, we talked about leading a series of workshops open to anyone within the organization that would be focused on identifying New Star's organizational values. To give them a place to start, we created an initial flow of a values workshop:

» Before the session, ask each participant to do some pre-work, selecting three to five of the values they most closely identify with the New Star culture (see the list of sample values is in the back of the book).
» At the beginning of the session, form small groups of no more than five people. Within the groups, give each person time to explain why they chose those specific values. Have the group pick three to five values they are aligned on.
» Ask one person from each group to share the values selected and why.
» Have each group post their values to a wall, where they're sorted for duplicates and similarities.

» Give the group time to review all the values on the wall and dot-vote on five values they feel best fits the organization.
» Facilitate a discussion to identify the top five values, digging into how everyone interprets the value and establishing definitions. Based on background and experiences, the meaning one person has of a value may be completely different for another. Make sure that everyone has a say; for the values to be meaningful, the workshop needs to be inclusive.

Jenny and Pete offered to coordinate the first values workshop and bring back the initial discoveries to the group in two weeks.

We then talked about how to integrate into interviews some questions that will help better understand if the candidate not only aligns with our culture but can make it better. It's a big ask. Hiring experts at LinkedIn, Airbnb, and other industry-leading organizations recommended a few questions that we can consider to help us get going:

» *How do your colleagues benefit from working with you specifically as opposed to a co-worker?*
This question helps us understand how the candidate sees themselves and how their unique skills and background can be of benefit to others. It also provides some insight into their thinking about collaboration.
» *Tell me about a time when understanding someone else's perspective helped you accomplish a task or resolve an issue?*
This question shows the candidate's ability to consider other ways of thinking and compare them with their own.

> *What is your impression of our company's culture and values?*
> This is a general question to help us understand how we could improve how we live and talk about our values. It also provides an objective perspective on New Star's culture and can help us see if the candidate is looking to fit in or bring about positive change.

The team agreed to consider these questions and how to incorporate them into the next set of interviews. They had a solid lineup of interviews scheduled and were getting more comfortable in how they could find that missing puzzle piece.

ACTIVITY: POST-INTERVIEW SELF-REFLECTION

After an interview, write down your own impression of the candidate with an eye toward looking for that missing puzzle piece that will push your teams to be better. Here are a few prompts to shape your thinking:

> » Will this candidate challenge the team's current thinking and process? What did you hear during the interview to support this?
> » Does the candidate bring new energy to the table? What made you feel that way?
> » Did you learn something new from the candidate? What was it?

Bring your thinking back to your hiring team, during a calibration session or other debrief opportunity. No matter if you feel the candidate will be that missing puzzle piece or another layer in the stack, explain where your thinking is coming from. Was it something the candidate said? Their body language? Give it detail and don't settle for "It's hard to describe" or "I can't put my finger on it." Understand, acknowledge, and stand behind your rationale.

Reflecting on and talking about these questions can also help you and your team become more aware of how inherent, unconscious biases are—even unintentionally—impacting your impressions of a new candidate. Unconscious biases can massively affect our ability to hire for diversity. Self-reflection and group dialogue are just small steps to building individual and organizational awareness of the existence of bias in hiring.

How bias impacts hiring is a deep topic that easily requires an entire book. If you'd like to read more, please check out the Reading List at the end of the book.

CHAPTER 6: ADOPTING CONTINUOUS LEARNING

"At most companies, people spend 2 percent of their time recruiting and 75 percent managing their recruiting mistakes."

—Capital One CEO, Richard Fairbank

I'm happy to say that after a lot of experimentation with the new structures the team found a candidate they felt was best suited to the new senior product manager role, and the candidate accepted the invitation to join New Star. However, the learning didn't stop there.

For many product organizations, including New Star, hiring has been considered a linear process that starts with a job description and ends with an offer, looking like a traditional sales funnel versus an approach to bringing new people on to your team.

TRADITIONAL HIRING PROCESS

Following a linear approach means missing out on opportunities to integrate learnings gathered throughout your hiring process. For example, during a calibration session for the new senior product manager role, a few of the team members realized that although they, as a team, had identified conflict resolution skills as priority for the role, in interviews, the candidates seemed taken aback by questions about how they deal with conflict. One team member went back to the job description and found that they forgot to include any mention of conflict resolution skills in the final job description. In a full-steam ahead linear hiring process, this wouldn't have been picked up on.

| JOB DESCRIPTION | CV SCREENING | PHONE INTERVIEW | IN-PERSON INTERVIEW | TAKE HOME ASSIGNMENT | HIRING CHAT | OFFER |

In fact, each item that we've discussed so far can be updated and refined based on quantitative and qualitative inputs. When reviewing CVs, did you notice that more men than women were applying? If so, should you consider changes to the language of the job description? When interviewing, did you realize that the people you were meeting had the technical skills listed in the job description but not many (if any) of the human skills that you think the role needs in order to be successful? If so, should you consider revisiting your role description to make sure both sets of skills are thought through and captured? As a cumulative, what do these insights say about your thinking on fit and diversity?

Just like a set of dominos, a change in one of these components can impact others—they're implicitly interconnected. As such, your hiring process needs to be a dynamic one that continually senses and responds to the evolving nature of finding the right person for the right role.

For those product people involved in hiring, think of it this way: In your product practice, would you ever think of building a product as a linear process? No. And, you shouldn't think of finding the right person for your role as a linear process either. Hiring is much more akin to a cycle of continual learning than a process.

HIRING FOR PRODUCT EQ: CONTINUAL LEARNING LOOP

Before New Star published its new senior product management role, the hiring team agreed to build new touch points into their hiring practices to better identify and respond to changes as they came up. The team scheduled retrospectives (aka "retros") throughout the hiring cycle—like the calibration sessions—to provide space for each member of the team to share what had gone well with the search since they last met, what hadn't gone so well, and what changes they may need to consider. The frequency of the meetings ebbed and flowed as needed; they tended to be weekly initially and then moved to biweekly.

Retros are a seemingly simplistic practice that any team can use to enable continuous learning throughout the hiring process, allowing the team to regroup, share, process learning, and collectively design change and decide how to make it actionable.

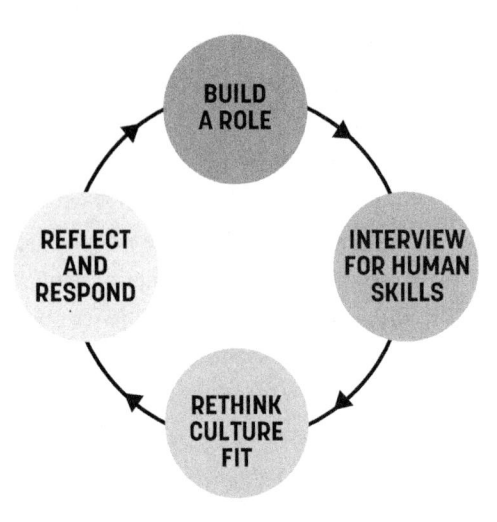

```
        BUILD
        A ROLE
   ↗              ↘
REFLECT          INTERVIEW
  AND            FOR HUMAN
RESPOND            SKILLS
   ↖              ↙
        RETHINK
        CULTURE
          FIT
```

While having a retro may seem like an easy answer, for many organizations, actually taking the time to stop and reflect just seems like a waste of time. There's often so much to do that stopping and actually thinking about the work and reflecting on it is the lowest priority. However, research by Giada Di Stefano, Francesca Gino, Gary Pisano, and Bradley Staats shows that deliberate reflection points—like retros—are key to organizational learning, and that individuals can perform up to 23 percent better after consistently reflecting on their work than they are by doing more work.

On the new senior product manager's first day, Pete added another retro to the calendar, this one scheduled for six months down the line and called a "role retro." Designed to ensure that the learning doesn't stop at onboarding, the role retro gives everyone involved in the hiring process—including the new hire herself—a chance to come back together and reflect on how the team did: Was the description of the role solid or did it miss that mark? Did the members of the hiring and interviewing crew make good decisions? What does your new team member think of the experience? Do they feel the role as described to them was an accurate description? What was missed? Did you make a good hiring decision?

The concept of a role retro is inspired by Jack Welch, former CEO of GE, who made hiring a priority for his teams and incentivized his organization to make good hiring decisions through a metric called the "Hiring Batting Average." Taken from the batting average common to baseball, it's a metric that measures how often a player hits the ball out of the total number of attempts at bat.

As described in Goleman's *Grow: Identifying and Fostering Talent*, Welch would ask each member if they were in favor of the hire or not. A year later, he went back to check in with the result of that decision. This simple metric helped to identify those who may not have been taking the hiring process seriously or were struggling to understand what might be a good decision for the team and organization.

As with all retrospectives—be it for a product or hiring process—taking time to step back, reflect, and learn gets us closer to our desired outcomes.

ACTIVITY: CONTINUOUS LEARNING WITH RETROSPECTIVES

Incorporate retrospectives into your hiring cycle from the beginning, making it part of your operating rhythm. It could be initially weekly as it was for New Star, or biweekly.

The goal of a retro is to create a safe space for group reflection and learning. There are countless formats to play with to help the team think about what's going well, what isn't going so well, and what needs to change in the hiring cycle. A few of the most popular are:

» Stop, Start, Continue: Actions and behaviors we should stop, start or continue doing

- » Good, Bad, Better, Best: Things that went well, didn't go well, opportunities for improvement, and things that deserve recognition
- » Continue (actions we want to continue) and Consider (actions to consider changing)

Between retros, make space for individual reflection. Setting aside 10 minutes during your day, even on your daily commute, to do a personal retro on the day builds your own muscle of learning through self-reflection and can increase performance, improve your mood, and decrease your risk of burnout.

CHAPTER 7: CREATING LASTING CHANGE

At the time of writing this book, I'm still working with New Star. Not because we haven't made progress, but as I said in my first meeting with Sarah and Pete, real change takes time, discipline, and dedication.

So far, we've made great strides in changing how Sarah, Pete, and the product teams think about product management—moving from a misunderstood domain to acknowledging it as a practice that demands both technical and human skills. We've radically changed what the organization thinks about when it comes to the skills that product people need and how hire to for them.

The team has scaled back their initial goal of immediately hiring 10 new people but have added two new team members to date. I'm happy to say that neither brainteaser interview questions nor job descriptions that were a product of cutting and pasting were involved. Instead, teams and stakeholders came together to build a role and interviews were thoughtfully prepared for and included a focus on human skills.

Thinking has shifted from hiring as a linear process to a cycle of continuous learning. We've created new structures to understand how to look for that missing puzzle piece, instead of adding another layer to the stack—making a conscious effort to build more diverse product teams.

This change hasn't been easy or fast, but it's happening. None of it would've been possible without Sarah, Pete, and members of the product team acknowledging that in order to create change of this magnitude they had to start with themselves. Because to hire for human skills you have to have them. If you're going to see them, you have to have them.

Pete, in particular, has made great strides. Through our coaching sessions and the on-going work with the product teams, Pete has continued to focus on building self-awareness. His initial outreach to team members for feedback has become a regular practice that has laid the foundation for better communication with team members. And, while Pete's inclination to micro-manage in times of stress hasn't been eliminated completely, members of the teams can more easily identify the pattern and actually surface concerns to Pete directly instead of sitting in the discomfort.

Most of all, Pete's focus on self-awareness has shown him that the technical skills that he relied on in the past to get him to this point in his career aren't the same skills that will take him forward. He's becoming an advocate of making human skills an essential part of the product practice at New Star and integrating new practices—like meeting check-ins and check-outs, and retros – into their ways of working.

I am so impressed with the effort and dedication the teams have put into developing their human skills, learning what it really means to be self-aware, adaptive, influence, and lead (among many other things). The benefits are showing-up in how they meet, how they communicate, how they think about the culture at New Star, and how and who they hire.

Just as we can learn new technical skills, our brains are also equipped to learn new human skills. Research from Goleman's now infamous *Harvard Business Review* article, "Leadership That Gets Results," and many others shows that it *is* possible to grow and improve our emotional intelligence and human skills. It is akin to changing a habit, so it takes more time and commitment as each habit must be unlearned and then replaced with a new one, but it is entirely feasible.

Opportunities for growth in any—and all—human skills are open to everyone, and the benefits are endless.

APPENDIX

HUMAN AND TECHNICAL SKILLS EXAMPLES

Use the following lists of human and technical skills for working through the exercises in this book. **These lists are not exhaustive** but are designed to spark your thinking about what makes up a balanced product practice.

Examples of Human Skills

This list includes examples of human skills based on my own firsthand experience with clients, and was supplemented by findings in The World Economic Forum's "Future of Jobs Report 2018," Adobe 99U's "Ten Human Skills for the Future of Work," and Goleman's "Emotional Intelligence has 12 Elements: Which do you need to work on?"

- » Accountability
- » Achievement orientation
- » Active listening
- » Active/Continuous learning
- » Adaptability
- » Coach and mentor
- » Collaboration
- » Complex problem solving
- » Conflict resolution
- » Creativity
- » Critical thinking
- » Curiosity
- » Decision making
- » Drive
- » Effective communication
- » Emotional intelligence (self-awareness, self-management, social awareness, relationship management)
- » Empathy
- » Influence
- » Initiative
- » Innovation
- » Leadership
- » Motivation
- » Organizational awareness
- » Originality
- » Resilience
- » Positive outlook
- » Team oriented

Examples of Technical Skills

- » A/B and multivariate testing
- » Agile
- » Budgeting and financial forecasts
- » Business case creation
- » Business model canvas
- » Competitive analysis
- » Customer interviews
- » Customer personas
- » Design sprints
- » Hypothesis-driven experimentation
- » Jobs-to-be-Done (JTBD)
- » Key-performance indicators (KPIs)
- » Lean
- » Minimum Viable Products (MVPs)
- » Objectives and key results (OKRs)
- » Portfolio management
- » Pricing
- » Product analytics
- » Product discovery
- » Product prioritization
- » Product prototypes
- » Product roadmaps
- » Product strategy
- » Qualitative user research
- » SEM and SEO
- » Scrum
- » Usability testing
- » User stories and acceptance criteria
- » Vision statements

Examples of Values: For use in values workshops

For a more extensive list of core values, go to https://www.threadsculture.com/core-values-examples.

- Accountability
- Adaptability
- Agility
- Authority
- Balance
- Belonging
- Boldness
- Community
- Cooperation
- Creativity
- Challenge
- Change
- Connection
- Continuous improvement
- Curiosity
- Customer focus
- Customer satisfaction
- Delight
- Dependability
- Differentiation
- Drive
- Discovery
- Education
- Efficiency
- Excellence
- Excitement
- Elegance
- Empowerment
- Enthusiasm
- Entrepreneurship
- Equality
- Flexibility
- Freedom
- Fun
- Growth
- Hard work
- Harmony
- Heart
- Heroism
- Honesty
- Humility
- Imagination
- Impact
- Inclusion
- Individuality
- Integrity
- Intuition
- Knowledge
- Leadership
- Listening
- Mastery
- Meaningful work
- Openness
- Optimism
- Originality

- Passion
- Partnership
- People
- Performance
- Personal development
- Positivity
- Pragmatism
- Profits
- Quality
- Relationships
- Reliability
- Resilience
- Respect
- Safety
- Speed
- Spontaneity
- Sustainability
- Teamwork
- Timelessness
- Tradition
- Transparency
- Trust
- Uniqueness
- Winning
- Wisdom
- Work-Life Balance

READING LIST

BUILDING A PRODUCT PRACTICE

For more on the human ability to improve EQ:

> Goleman, Daniel. "Leadership That Gets Results." *Harvard Business Review*, Mar.-Apr. 2000.

UNDERSTANDING EQ AND HUMAN SKILLS

The definition, history, theories, and frameworks related to emotional intelligence came from:

> Institute for Health and Human Potential. "What Is Emotional Intelligence?" https://www.ihhp.com/meaning-of-emotional-intelligence.

> Riopel, Leslie. "Goleman and Other Key Names in Emotional Intelligence Research." PositivePsychology.com. https://positivepsychology.com/emotional-intelligence-goleman-research/.

> Goleman, Daniel and Boyatzis, Richard E. "Emotional Intelligence Has 12 Elements: Which Do You Need to Work On?" *Harvard Business Review*, 06 Feb. 2017. https://hbr.org/2017/02/emotional-intelligence-has-12-elements-which-do-you-need-to-work-on.

> Big Think Edge. "Daniel Goleman Introduces Emotional Intelligence," *YouTube*, 23 Apr. 2012. https://www.youtube.com/watch?v=Y7m9eNoB3NU.

For more specifically on human skills and Product EQ, visit my blog https://www.KateLeto.com/.

DECIPHERING THE JOB DESCRIPTION

Survey data that 50 percent of job descriptions are the result of cutting and pasting comes from:

> Spencer, Patrick. "3.75 Million Labor Hours Spent Rewriting Job Descriptions Annually." *The Mighty*

Recruiter, 15 Aug. 2016. https://www.mightyrecruiter.com/blog/3-million-hours-writing-job-descriptions/.

USING THE ROLE CANVAS
The Role Canvas was initially inspired by work with The Ready, an organizational design and transformation firm: https://theready.com/.

Fundamental thinking on creating roles comes from:

> Dignan, Aaron. *Brave New Work*. Penguin Random House UK, 2019.

INTERVIEWING FOR HUMAN SKILLS
The research behind behavior-based interview questions and specific examples comes from:

> Lynn, Adele B. *The EQ Interview: Finding Employees With High Emotional Intelligence*. American Management Association, 2008.

Read more about the practice of check-ins and check-outs here:

> Gerber, Niklaus. Medium. April 17, 2020. "How Check-Ins and Check-Outs will help you to build stronger teams." https://medium.com/@niklausgerber/team-check-ins-and-check-outs-376aaef9357f

GOING BEYOND CULTURE FIT

> "Ten Ways of Managing Person-Organization Fit (P-O Fit) Effectively: A Literature Study," *International Journal of*

> *Business and Social Science*, Vol. 2 No. 21, Special Issue – Nov. 2011. http://ijbssnet.com/journals/Vol_2_No_21 _Special_Issue_November_2011/25.pdf.

> Hewlett, Sylvia Ann, et. al. "How Diversity Can Drive Innovation," *Harvard Business Review*, Dec. 2013. https://hbr.org/2013/12/how-diversity-can-drive-innovation.

For more about culture complements, check out:

> Vijungco, Jeff. "Don't Fit In," Linked In, 29 Mar. 2017. https://www.linkedin.com/pulse/dont-fit-jeff-vijungco/.

The interview questions to help assess culture add came from:

> "How to Assess for Culture Add," Linked In Talent Solutions. https://business.linkedin.com/talent-solutions/recruiting-tips/how-to-assess-skills/culture-add.

ADOPTING CONTINUOUS LEARNING

Research into the relationship between reflection and performance at work came from:

> Di Stefano, Giada, et. al. "Making Experience Count: The Role of Reflection in Individual Learning." Harvard Business School. https://papers.ssrn.com/sol3/papers.cfm?abstract_id=2414478##

> Goleman, Daniel, et. al. *Grow: Identifying and Fostering Talent*. More Than Sound, 2015.

More detail on the retrospectives listed in the activity can be found here:

> Gratis, Brandi. "3 Popular Ways to Run a Productive Retrospective." Backlog, 14. Nov. 2016. https://backlog.com/blog/three-ways-run-productive-retrospective/.

FURTHER READING

Here are a few other interesting titles to continue your learning:

Agarwal, Pragya. *Sway: Unravelling Unconscious Bias.* Bloomsbury Sigma, 2020.

Agarwal, Pragya. "Here Is How Bias Can Affect Recruitment in Your Organization." Forbes, 19 Oct. 2018. https://www.forbes.com/sites/pragyaagarwaleurope/2018/10/19/how-can-bias-during-interviews-affect-recruitment-in-your-organisation/#eb3123f1951a

Bradberry, Travis and Jean Greaves. *Emotional Intelligence 2.0.* Talent Smart, 2009.

Eurich, Tricia. *Insight: The Power of Self Awareness in a Self-Deluded World. Currency, 2017.*

Goleman, Daniel. *Emotional Intelligence: Why It Can Matter More Than IQ.* Bantam, 2006.

Kahneman, Daniel. *Thinking Fast and Slow.* Farrar, Straus and Giroux, 2013.

Little, Jason. *Lean Change Management: Innovative Practices for Managing Organizational Change.* Happy Melly Express, 2014.

Madden, Debbie. *Hire Women: An Agile Framework for Hiring and Retaining Women in Technology.* Sense & Respond Press, 2018.

Thomas, Adam. "Hiring for Product Is Broken." Mind the Product, 11 June 2020. https://www.mindtheproduct.com/bias-in-interviews-and-why-hiring-in-product-is-broken/.

For more information on all of the activities featured in the book, please visit my blog at https://www.kateleto.com/

ACKNOWLEDGMENTS

This is my first foray into writing a book and as with doing anything for the first time, there's been a learning curve! This final product wouldn't have been possible without the support and guidance of many people. Just to name a few, I'd like to thank: Faye Benfield, Martin Eriksson, Adrienne Tan and Mike MacIntyre, all early readers who provided valuable product perspective. Wayne Palmer and Shereen Hoban, both coaches that I've worked with who reviewed various versions of this book and were generous with their time and advice. Jeff Gothelf and Josh Seiden and the team at Sense & Respond Press for making this opportunity possible and encouraging me along the way. Kathryn Maloney, my friend and colleague who has listened to me talk about this project for a long time and always been patient, reflective, and supportive. Sharada Thompson who has helped me personally to become open to the countless benefits of "human skills"; it truly has been a beautiful journey, Sharada. And to Neil (of course, of course!)—your support and patience has made all the difference.

KATE LETO's product management, org design, and marketing background spans more than 25 years. She has had a front-row seat to the evolving ways products are discovered, defined, built, and delivered and now takes her hands-on experience into organizations of all shapes and sizes as a consultant, coach, and advisor; helping to create authentic, high-performing cultures, teams, and products. Her consulting experience has taken her around the world, guiding clients that range from disruptive startups to Fortune 500 companies.

www.kateleto.com

Printed in Great Britain
by Amazon